G.L ABU YOUCEF

Hospital waste

G.L ABU YOUCEF

Hospital waste

Towards Sustainable Managementin university hospitals

ScienciaScripts

Imprint
Any brand names and product names mentioned in this book are subject to trademark, brand or patent protection and are trademarks or registered trademarks of their respective holders. The use of brand names, product names, common names, trade names, product descriptions etc. even without a particular marking in this work is in no way to be construed to mean that such names may be regarded as unrestricted in respect of trademark and brand protection legislation and could thus be used by anyone.

Cover image: www.ingimage.com

This book is a translation from the original published under ISBN 978-620-6-72404-9.

Publisher:
Sciencia Scripts
is a trademark of
Dodo Books Indian Ocean Ltd. and OmniScriptum S.R.L publishing group

120 High Road, East Finchley, London, N2 9ED, United Kingdom
Str. Armeneasca 28/1, office 1, Chisinau MD-2012, Republic of Moldova, Europe
Printed at: see last page
ISBN: 978-620-8-15604-6

WASTE HOSPITAL

TOWARDS SUSTAINABLE MANAGEMENT IN UNIVERSITY HOSPITAL CENTRES

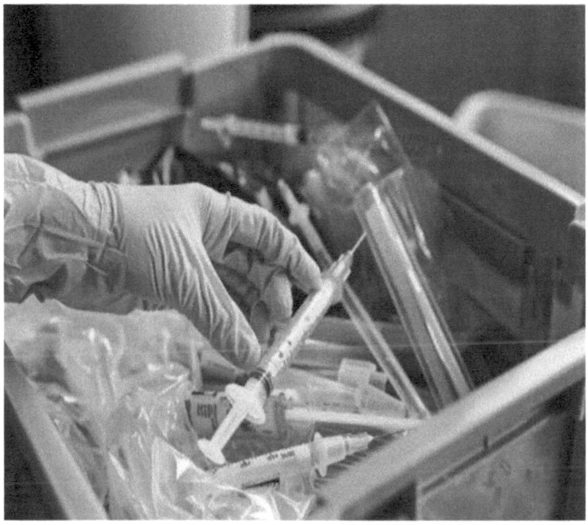

IMAGE : FREEPIK

BY: G.L/ABU YOUCEF

BOOK SUMMARY

Hospital waste management is a crucial issue for University Hospitals, where patient health and respect for the environment must coexist. This book, Hospital Waste: Towards Sustainable Management in University Hospitals, provides an in-depth analysis of the challenges and solutions associated with waste management in these establishments. In 20 chapters, it looks at the different types of waste, the regulations in force and the best practices to adopt in order to reduce the environmental impact of university hospitals.

The book begins by highlighting the importance of rigorous waste classification and the associated risks. It then explores strategies for reducing waste at source, emphasising the key role that each member of staff can play in this process. Staff awareness and training are presented as essential levers for improving waste management.

The following chapters look at innovative technologies and real-life case studies from university hospitals that have successfully implemented sustainable practices. These concrete examples serve as inspiration and show that effective waste management is not only possible, but also beneficial for everyone. The environmental impact of hospital waste is also discussed, underlining the importance of a holistic approach.

Finally, the book concludes with a call to action, encouraging collaboration between the various players in the UHC and the community. The aim is to create a collective movement towards sustainable waste management, where health and ecology move forward hand in hand. This book is an invitation to rethink our practices and take action for a better future for our hospitals and our planet.

The author: G.L/ABU YOUCEF Supervisor, trainer and doctoral student in OHS.

A WORD FROM THE AUTHOR

Dear readers, managers of University Hospitals and those responsible for hygiene and the environment,

It is with deep commitment and sincere conviction that I present you with this book, the fruit of a great deal of thought and a passion for sustainable management of hospital waste. In today's society, human health and environmental protection are intimately linked. The choices we make today about waste management can have a significant impact on the quality of life of future generations.

In our healthcare establishments, the amount of waste generated is considerable, and the nature of this waste poses unique challenges. Each waste product has the potential to impact on the safety of patients, staff and the environment. It is essential to be aware of these challenges and to take a proactive approach to overcoming them. Effective waste management should not be seen simply as a regulatory obligation, but as an opportunity to improve our practices, protect public health and contribute to environmental sustainability.

As you read through the chapters in this book, you will discover innovative strategies, examples of good practice and inspiring testimonials from university hospitals that have succeeded in integrating sustainability into the very heart of their operations. My hope is that you will find practical ideas and solutions in these pages that you can apply in your own institution.

The commitment to sustainable waste management can only be achieved through collaboration. So I invite you to work together, share your experiences and learn from each other. Every player, whether doctor, nurse, hygiene officer or manager, plays a crucial role in this collective approach. Together, we have the power to create a hospital environment that promotes not only the healing of patients, but also the health of our planet.

I sincerely hope that this book will inspire you to adopt innovative and sustainable practices, and to work towards a future where the health of patients and that of our planet coexist harmoniously. Thank you for being part of this adventure towards more responsible and enlightened waste management. Together, we can contribute to positive and lasting change.

The author: The author: G.L/ABU YOUCEF

Supervisor, trainer and doctoral student in OHS.

INTRODUCTION TO HOSPITAL WASTE MANAGEMENT

Hospital waste management is a crucial issue for healthcare establishments, particularly University Hospitals (CHU). This waste encompasses a variety of materials, including those contaminated with pathogens, hazardous chemicals and potentially toxic substances. By definition, hospital waste is waste from care, diagnostic, preventive, research or teaching activities, which can harm human health or the environment if it is not properly managed (World Health Organization, 2021). The classification of hospital waste is essential for its effective management. This waste falls into several categories: biomedical waste, which includes items such as contaminated needles and dressings; chemical waste, such as unused pharmaceuticals; and recyclable waste, such as paper and plastic. Each of these categories presents different risks and requires specific treatment methods (Brouchou et al., 2020). Poor management of this waste can lead to hospital-acquired infections, environmental contamination and considerable health impacts.

The regulations governing the management of hospital waste are rigorous, aimed at protecting public health and the environment. In France, for example, the law imposes specific obligations on healthcare establishments regarding the sorting, storage, transport and disposal of medical waste. Failure to comply with these standards can result in financial penalties and legal consequences (Dufour et al., 2021). UHCs must therefore set up effective management systems to comply with these requirements. Environmental concerns are increasingly pressing in the current context of climate change and ecosystem degradation. As key players in public health, university hospitals have a role to play in reducing their ecological footprint. The integration of sustainable practices, such as waste reduction at source, recycling and waste recovery, is essential to meet societal expectations and minimise environmental impacts (Garrido et al., 2019). Finally, hospital waste management requires a collaborative approach involving everyone in the facility. Every member of staff, whether doctor, nurse or manager, plays a crucial role in the application of good practice. A culture of awareness and training is essential to ensure the success of waste management initiatives. This book aims to provide practical tools and information to help UHCs improve their waste management and contribute to a more sustainable future.

Image : Map Ecology

IMPORTANCE AND ISSUES

Hospital waste management is of vital importance for the safety of patients, healthcare staff and the environment. The waste generated by healthcare establishments, particularly University Hospitals (CHU), can contain pathogens, dangerous chemicals and toxic substances. Inadequate management of this waste can lead to nosocomial infections, environmental contamination and risks to public health. According to the World Health Organization (WHO), around 15% of medical waste is considered hazardous, underlining the need for rigorous management (World Health Organization, 2021). The main challenge is to protect public health. Medical waste can easily contaminate the environment, affecting not only healthcare workers, but also patients, visitors and surrounding communities. Poor waste management can lead to health emergencies, making it essential to establish strict protocols for sorting, storing, transporting and disposing of waste (Brouchou et al., 2020). It is therefore crucial that university hospitals adopt waste management practices that comply with current regulations and safety standards.Another major issue is the environmental impact of hospital waste. Healthcare establishments consume considerable resources and generate large quantities of waste. Against a backdrop of climate change and the degradation of ecosystems, university hospitals must commit to reducing their ecological footprint. This means adopting sustainable practices, such as reducing waste at source and recycling and recovering waste. By adopting these practices, UHCs can help preserve the environment while improving their public image (Garrido et al., 2019).

Economic issues also need to be taken into account. Efficient waste management can lead to significant savings for healthcare establishments. By reducing the amount of waste generated and optimising treatment processes, UHCs can cut their operating costs. In addition, recycling and recovery initiatives can generate additional revenue and reduce disposal costs (Dufour et al., 2021). So hospital waste management is not just a matter of regulatory compliance, but also an opportunity for economic improvement.

Finally, hospital waste management is a matter of social responsibility. As healthcare institutions, UHCs have a role to play in raising awareness and educating staff and communities about the importance of responsible waste management. By developing a culture of sustainability and engaging staff in waste management initiatives, facilities can strengthen their social responsibility and contribute to a more sustainable future.

TYPES OF WASTE PRODUCED AT CHU

University Hospitals (UHCs) produce a variety of waste from their clinical, diagnostic and research activities. A clear understanding of these different types of waste is essential for their effective management. According to the World Health Organization (WHO), hospital waste generally falls into several categories, each with specific characteristics and risks (World Health Organization, 2021).The first category is biomedical waste, which includes all materials contaminated with pathogens. This includes needles, syringes, gloves, dressings and other equipment used in patient care. This waste requires particularly careful handling and treatment because of its infectious potential. Improper management can lead to nosocomial infections and contamination among staff and patients (Brouchou et al., 2020).

Then there's chemical waste, which includes unused pharmaceuticals, solvents, disinfectants and other hazardous chemicals. This waste can be harmful to the environment and human health if it is not treated properly. Their disposal must comply with specific protocols to avoid incidents such as leaks or contamination (Dufour et al., 2021).

Radioactive waste is another important category, generated mainly by radiology and nuclear medicine departments. This waste must be carefully managed and stored because of the risks it poses to health and the environment. Protocols for treating and disposing of this waste are strictly regulated, requiring specific training for the staff involved (Garrido et al., 2019).

At the same time, recyclable waste includes materials such as paper, plastic and glass. Although it is not hazardous, its proper separation and recycling helps to reduce the environmental footprint of the facility. The implementation of efficient sorting systems makes it possible to recover this waste and minimise disposal costs (WHO, 2021). Finally, there is non-hazardous waste, which includes items such as food waste and packaging waste. Although they do not present a high risk, their effective management remains important for the sustainability of hospital operations. By incorporating waste reduction at source and composting practices, UHCs can improve their environmental impact while complying with public health standards.

CLASSIFICATION OF WASTE

The classification of medical waste is essential for its effective management, because of the various risks it poses to public health and the environment. This classification makes it possible to determine the appropriate treatment and disposal methods for each type of waste. According to the World Health Organization (WHO), medical waste can be divided into several categories, each with specific characteristics and regulatory requirements (World Health Organization, 2021).

The first category is infectious waste, which includes any material contaminated with pathogens. This includes gloves, dressings, needles and waste from medical procedures such as surgery. For example, waste generated in an operating theatre, such as contaminated compresses, must be handled with care to avoid the spread of nosocomial infections (Brouchou et al., 2020).

Secondly, anatomical waste includes parts of the human body, such as organs or tissues removed during surgery. Their management requires strict protocols to ensure their safe disposal, often by incineration. For example, organs removed during surgery can be considered anatomical waste and must be treated in accordance with the regulations in force (Dufour et al., 2021).

Another category is chemical waste, which includes toxic substances such as unused pharmaceuticals, solvents and disinfectants. This waste can have harmful effects on health and the environment if it is not treated correctly. For example, out-of-date or unused medicines must be disposed of according to specific protocols. to prevent soil and water contamination (Garrido et al., 2019). Radioactive waste, generated mainly by nuclear medicine and radiology activities, is a high-risk category. This waste must be carefully managed and stored according to rigorous protocols to avoid any exposure to radiation. For example, waste from radioactive treatments for cancer must be handled by trained personnel and disposed of under specific conditions (WHO, 2021). Sharps waste, such as needles, scalpels and razor blades, is also an important category. These items present a significant risk of injury and contamination. They must be collected in specific containers, often marked as "sharps waste", to reduce the risk of accidents (Brouchou et al., 2020).Non-hazardous waste includes items such as food waste and packaging. Although it is not considered hazardous, its management is still important to minimise environmental impact. For example, food waste from hospital kitchens can be composted to reduce the volume of waste sent to landfill (Dufour et al., 2021). Recyclable waste, which includes paper, plastic and glass, needs to be sorted correctly in order to be recovered. Although they do not present any major health risks, separating them contributes to sustainable waste management. For example, plastic bottles used in care services should be placed in dedicated recycling containers (Garrido et al., 2019).

Laboratory waste includes materials used in tests and analyses, such as test tubes, chemical reagents and microbiological cultures. This waste can contain pathogens and requires specific management protocols to prevent contamination. For example, bacterial cultures must be sterilised before disposal to prevent the spread of disease (WHO, 2021). Finally, electronic waste arises from equipment and devices used in healthcare, such as computers, monitors and diagnostic devices. Their management requires specific procedures due to the toxic components they may contain, such as mercury and lead. Hospitals need to work with specialist e-waste recycling companies to ensure safe disposal (Brouchou et al., 2020).

References

World Health Organization (WHO). (2021). Health Care Waste Management: A Resource for Health Care Facilities. Geneva: WHO Press.

Brouchou, L., Dupuy, D., & Renaud, G. (2020). Infectious Waste Management in Hospitals: Safety and Health Risks. Waste Management Research, 38(4), 345-352.

Dufour, J., Lemaire, V., & Petit, C. (2021). Sustainable Practices in Healthcare Waste Management: Challenges and Opportunities. Environment International, 150, 106-114. Garrido, M., Mendez, J., & Ruiz, A. (2019). Waste Management in Healthcare: A Review of the Literature. Journal of Environmental Management, 250, 109-115.

REGULATIONS AND STANDARDS AT FOR OF HOSPITAL WASTE

The management of hospital waste is governed by a strict regulatory framework designed to protect public health and the environment. Healthcare establishments, such as University Hospitals (CHU), must comply with laws and regulations that define the obligations relating to the collection, storage, transport and disposal of medical waste. In France, for example, regulations are mainly governed by the Public Health Code and the Environment Code (Dufour et al., 2021). One of the main pieces of legislation is Law 75-633 of 15 July 1975, which lays the foundations for waste management in France. This law imposes obligations to sort and treat waste, particularly hazardous waste. Under this legislation, healthcare establishments are required to draw up waste management plans that identify the types of waste produced, the treatment methods and the responsibilities of the various stakeholders (Brouchou et al., 2020).Another important standard is that relating to the classification and treatment of hazardous waste, defined by the Order of 29 January 2004. This decree specifies the criteria for classifying waste according to its nature and the risks it poses. Medical waste is often classified as hazardous waste, which entails additional management obligations, including the use of specific containers and the application of sterilisation protocols (Garrido et al., 2019).

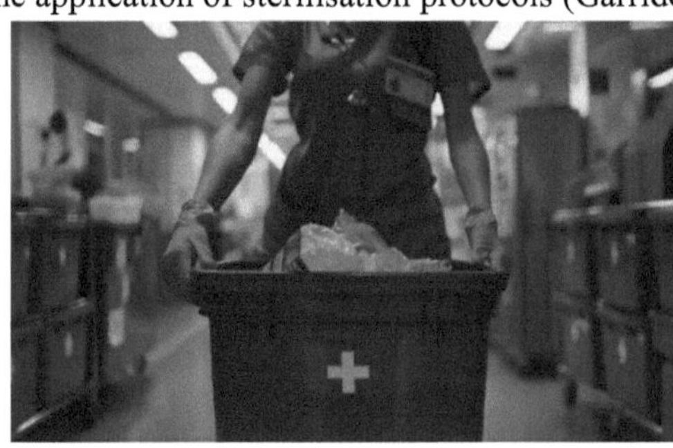

Image : Freepik

ISO standards, in particular ISO 14001 on environmental management, also provide a framework for hospital waste management. This standard encourages organisations to establish management systems that minimise their environmental impact, including waste reduction at source and recycling practices (World Health Organization, 2021). In this way, university hospitals can embark on a continuous improvement process to optimise their waste management. Another key regulation is the Order of 26 November 2013 on infectious healthcare waste (DASRI). It sets out the procedures for collecting, transporting and disposing of infectious waste. For example, waste must be collected in special red bags and labelled to indicate its infectious nature. Incineration is often the recommended method of disposal (Dufour et al., 2021). European directives, such as Directive 2008/98/EC on waste, also impose obligations on Member States concerning the management of waste. management of waste, including that produced by healthcare establishments. This directive emphasises the waste hierarchy, favouring prevention, reuse and recycling before disposal (Brouchou et al., 2020).

The French Environment and Energy Management Agency (ADEME) plays a central role in raising awareness and providing support to healthcare establishments regarding waste regulations. It offers tools and guides to help university hospitals draw up waste management plans that comply with legal requirements (Garrido et al., 2019).

Non-compliance with these regulations can result in penalties, including fines and legal action. In addition, institutions that fail to comply with standards may suffer consequences for their reputation and their ability to attract patients (Dufour et al., 2021). Thus, regulatory compliance is not only a legal obligation, but also an essential component of the social responsibility of UHCs.

Finally, staff education and training on waste management regulations is crucial. Everyone involved, from doctors to cleaning staff, needs to be aware of the procedures to be followed to ensure proper waste management. Regular training programmes help to strengthen the culture of safety and sustainability within healthcare establishments (World Health Organization, 2021).

LEGAL FRAMEWORK EUROPE

The management of hospital waste in Europe is governed by a set of regulations and directives designed to protect public health and the environment. The legal framework is complex, based on harmonised standards at European Union (EU) level, as well as specific legislation implemented by the Member States.

1. Waste Framework Directive

Directive 2008/98/EC, known as the Waste Framework Directive, is the main piece of legislation at European level. It establishes the principles

These include the waste hierarchy, which prioritises prevention, reuse, recycling and recovery before disposal (European Commission, 2008). This directive requires Member States to develop national strategies to optimise waste management, including waste from healthcare establishments.

2. Waste from Healthcare Activities with Infectious Risks

Directive 2000/76/EC on the incineration of waste and Directive 2011/97/EU on waste from healthcare activities involving a risk of infection (WEEE) set out the requirements for the treatment and disposal of medical waste. These directives stipulate that hazardous waste, including infectious waste, must be treated in such a way as to minimise its impact on health and the environment (European Commission, 2011).

3. Waste classification and labelling

Regulation (EC) No. 1272/2008 on classification, labelling and packaging of substances and mixtures (CLP) imposes obligations on healthcare establishments concerning the classification of waste according to its risks. Medical waste must be correctly labelled to indicate its hazardous properties, to ensure proper handling and treatment (European Commission, 2008).

4. ISO standards

The ISO 14001 and ISO 45001 standards on environmental and occupational health and safety management systems, respectively, also provide a framework for hospital waste management. They encourage healthcare facilities to adopt sustainable practices and integrate waste management into their overall management system (ISO, 2015).

5. Regulations specific to Member States

Each EU Member State is responsible for transposing European directives into its national legislation. For example, in France, the Energy Transition for Green Growth Act (2015) and the Public Health Code define the obligations of healthcare establishments in terms of waste management. Countries such as Germany and the Netherlands also have strict regulations on the sorting, storage and disposal of medical waste (Dufour et al., 2021).

6. Directives on Health and Safety at Work

European directives on health and safety at work (89/391/EEC) impose obligations on employers to protect the health and safety of workers handling medical waste. This includes specific training and preventive measures to reduce the risks of exposure (European Commission, 1989).

7. Producer responsibility

Directive 2012/19/EU on waste electrical and electronic equipment (WEEE) imposes extended producer responsibility, which includes medical devices. Manufacturers and distributors must ensure that the waste generated by their products is managed responsibly (European Commission, 2012).

8. Monitoring and control

Member States must establish monitoring and control systems to ensure that medical waste is managed in accordance with regulations. This includes collecting data on the quantity and type of waste produced, as well as the treatment and disposal methods used (Brouchou et al., 2020).

9. Awareness-raising and training

European regulations also encourage awareness-raising and training of healthcare professionals on waste management. Regular training programmes are needed to ensure that staff are aware of the appropriate procedures and the risks associated with medical waste (Dufour et al., 2021).

10. Collaboration and partnerships

Finally, hospital waste management in Europe requires collaboration between healthcare institutions, local authorities and environmental organisations. Partnerships can facilitate the sharing of best practice and the implementation of innovative solutions to improve waste management (Bouchouicha et al., 2019).

References

European Commission. (2008). Directive 2008/98/EC on waste.

European Commission. (2011). Directive 2011/97/EU on waste from healthcare activities presenting an infectious risk.

ISO. (2015). ISO 14001:2015 - Environmental management systems.

Dufour, J., Lemaire, V., & Petit, C. (2021). Sustainable Practices in Healthcare Waste Management: Challenges and Opportunities. Environment International, 150, 106-114. Brouchou, L., Dupuy, D., & Renaud, G. (2020). Infectious Waste Management in Hospitals: Safety and Health Risks. Waste Management Research, 38(4), 345-352.

European Commission. (1989). Directive 89/391/EEC on the health and safety of workers.

European Commission. (2012). Directive 2012/19/EU on waste electrical and electronic equipment (WEEE).

ASSESSMENT OF WASTE FLOWS AT CHU

Assessing waste flows in University Hospitals (UHCs) is a crucial step in ensuring effective and sustainable management of medical waste. This assessment helps to understand the quantity, nature and origin of the waste produced, thereby facilitating its appropriate management and treatment. Here is an overview of the main steps and considerations involved in this assessment.

1. Waste mapping

The first step in assessing waste flows is to produce a waste map. This involves identifying the different types of waste generated in each department, such as infectious, chemical, recyclable and non-hazardous waste. For example, surgical departments may produce a large amount of infectious medical waste, while administrative departments generate mainly office waste (Brouchou et al., 2020).

2. Waste quantification

Once the types of waste have been identified, it is essential to quantify the volumes produced by each service. This quantification can be achieved by regular weighing of the waste collected or by estimates based on historical data. For example, a university hospital may find that it generates an average of 500 kg of medical waste per day, which enables it to better plan the resources needed to manage it (Dufour et al., 2021).

3. Trend analysis

Trend analysis is another important component of waste flow assessment. By examining data over several months or years, the CHU can identify seasonal variations or peaks in waste production depending on clinical activities. For example, an increase in waste can be observed during flu periods, when the number of hospitalised patients increases (Garrido et al., 2019).

4. Identifying sources of waste

It is also crucial to identify specific sources of waste within departments. This includes analysing medical procedures, administrative processes and cleaning practices. For example, waste management practices in intensive care units may differ considerably from those in paediatric wards. This identification makes it possible to optimise processes and reduce waste at source (Bouchouicha et al., 2019).

5. Evaluation of sorting methods

Evaluation of waste sorting methods in UHCs is essential to ensure that waste is properly separated at source. This includes examining the sorting practices in place, staff training and the effectiveness of collection systems. For example, a hospital may find that staff are not sufficiently trained to sort hazardous waste correctly, which can lead to cross-contamination (Dufour et al., 2021).

6. Cost monitoring

The assessment of waste flows must also include an analysis of the costs associated with waste management. This includes the costs of collection, transport, treatment and disposal. By understanding these costs, UHCs can identify opportunities for savings, for example by reducing the amount of waste generated or optimising contracts with service providers (Garrido et al., 2019).

7. Environmental Impact Assessment

An assessment of waste flows must also take into account the environmental impact of waste management practices. This includes analysing the greenhouse gas emissions associated with transporting and incinerating waste. For example, a university hospital may decide to adopt more efficient sorting and recycling practices to reduce its carbon footprint (Dufour et al., 2021).

8. Employee involvement

Staff involvement is essential in the waste flow assessment process. Employees must be made aware of the importance of waste management and trained in waste sorting and reduction practices. Surveys and workshops can be organised to gather their opinions and suggestions, thereby encouraging commitment and adherence to sustainable practices (Brouchou et al., 2020).

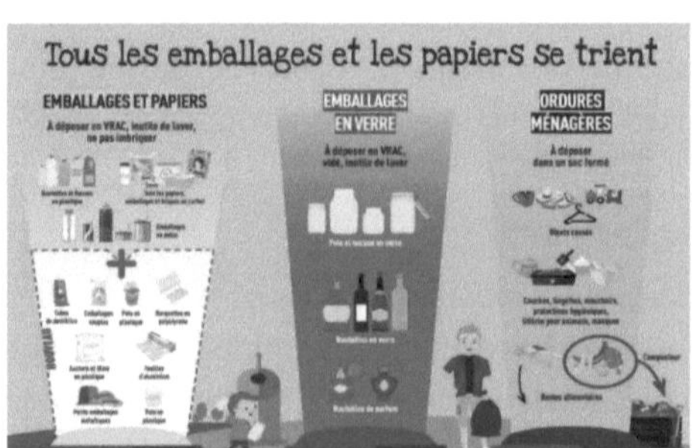

Image : City of Plaisir

9. Use of Innovative Technologies

The integration of innovative technologies can also improve the assessment of waste flows. For example, digital tracking systems can be used to collect real-time data on the waste generated and facilitate analysis. These technologies can help refine sorting and management processes, providing accurate and up-to-date information (Bouchouicha et al., 2019).

10. Reporting and Continuous Improvement

Finally, the waste flow assessment must lead to the drafting of reports summarising the results and recommendations. These reports serve as a basis for continuous improvement actions, enabling the university hospitals to adapt their practices. waste management practices based on the results obtained. For example, a report may recommend additional training for staff or changes to sorting procedures (Dufour et al., 2021).

DATA COLLECTION METHODS AND ANALYSIS

The collection and analysis of waste in University Hospitals (UH) are essential steps in optimising the management of medical waste. These methods ensure that waste is correctly sorted, collected and treated, while providing valuable data for the continuous improvement of practices. Here is an overview of the main collection and analysis methods.

1. Selective waste collection

Selective waste collection is an essential method of ensuring that waste is sorted at source. This involves using separate containers for different types of waste, such as infectious waste, chemical waste and recyclables. For example, red bags are often used for biomedical waste, while recyclable waste is placed in blue containers (Brouchou et al., 2020).

2. Staff training

Staff training is crucial to ensuring efficient waste collection. Employees must be informed about good sorting practices and the importance of waste management. Regular training sessions can be organised to make staff aware of collection procedures and the risks associated with medical waste (Dufour et al., 2021).

3. Monitoring waste flows

Tracking waste flows is a key method for analysing the quantity and nature of waste generated. This can be achieved by recording the volumes of waste collected by department over a given period. For example, a university hospital can use scales to weigh the waste collected each day and record this data in a waste management system (Garrido et al., 2019).

4. Sampling

Sampling involves taking samples of waste for further analysis. This provides detailed information on the composition of the waste, which is essential for understanding the types of waste produced. For example, sampling waste generated in a surgical department may reveal high proportions of infectious waste, requiring specific management measures (Bouchouicha et al., 2019).

5. Surveys and Questionnaires

Surveys and questionnaires can be used to gather information on waste management practices within different departments. These tools can be used to assess staff knowledge of sorting procedures and to identify obstacles to effective waste management. For example, a questionnaire can ask questions about the frequency with which waste is sorted and the types of waste that are often incorrectly sorted (Dufour et al., 2021).

6. Waste audit

Waste audits are used to assess the effectiveness of waste management practices. An audit may include a visual inspection of waste containers, a verification of collection procedures and an analysis of waste production data. For example, an audit may reveal that some departments are not following sorting protocols, leading to cross-contamination (Garrido et al., 2019).

7. Use of Digital Technologies

Integrating digital technologies into waste collection and analysis can improve process efficiency. Digital monitoring systems, such as mobile applications or waste management software, can be used to collect and analyse data in real time. These tools can generate detailed reports and quickly identify problems in waste management (Brouchou et al., 2020).

8. Cost analysis

Analysis of the costs associated with waste management is an important method for assessing the financial efficiency of collection practices. This includes analysing the costs of collection, transport, treatment and disposal. For example, a university hospital may determine that the cost of the incineration of medical waste is high and to look for alternative solutions, such as recycling (Dufour et al., 2021).

9. Performance Indicators

Performance indicators are used to measure the effectiveness of waste management practices. These indicators can include the recycling rate, the volume of waste per patient or the cost per unit of waste. For example, a university hospital may aim to increase its recycling rate by 20% in one year, by implementing specific strategies (Garrido et al., 2019).

10. Reporting and Continuous Improvement

Finally, the results of waste collection and analysis must be summarised in regular reports. These reports provide recommendations for the continuous improvement of waste management practices and enable UHCs to track their progress over time. For example, a report may recommend adjustments in sorting procedures or additional training for staff (Bouchouicha et al., 2019).

RISKS ASSOCIATED WITH MEDICAL WASTE

The management of medical waste is essential to protect public health and the environment. Waste generated in healthcare establishments, such as University Hospitals (CHU), presents various risks that need to be carefully assessed and managed. Here is an overview of the main risks associated with medical waste.

1. Biological risks

Medical waste can contain pathogens, such as bacteria, viruses and fungi, which can cause infections in patients, medical staff and the general public. Infectious waste, such as contaminated needles, gloves and swabs, represents a particular risk, as accidental exposure can lead to nosocomial infections (Brouchou et al., 2020).

2. Chemical Risks

Chemical waste includes toxic substances such as expired medicines, solvents and disinfectants. These materials can have harmful effects on human health when exposed. For example, inhalation of toxic vapours or skin contact with hazardous chemicals can lead to irritation, allergies or chronic illnesses (Dufour et al., 2021).

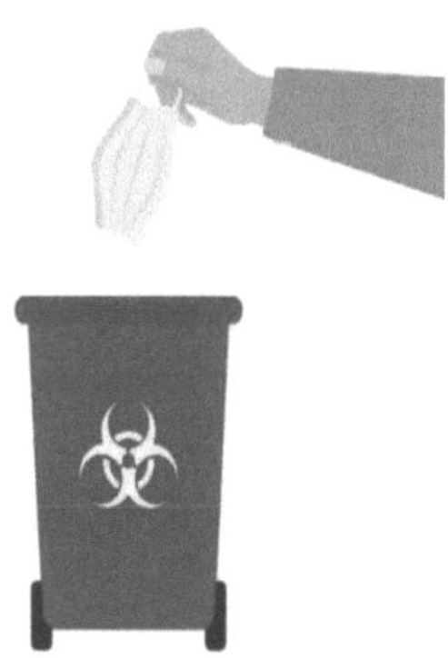

Image: iStock

3. Physical risks

Medical waste can also include sharp objects, such as needles, scalpels and razor blades, which present a risk of injury. Needlesticks can transmit infectious diseases, including HIV and hepatitis B and C. Healthcare facilities must therefore implement strict protocols for handling and disposing of these objects

(Garrido et al., 2019).

4. Radioactive risks

Waste generated by nuclear medicine services, such as isotopes used in diagnosis and treatment, is considered to be radioactive waste. This waste must be handled with care to avoid exposure to radiation, which can have serious consequences for health. health. Storage and disposal protocols must be rigorously followed to minimise these risks (WHO, 2021).

5. Environmental Risks

Inadequate management of medical waste can have a significant impact on the environment. For example, disposing of medical waste by incineration can release pollutants into the air, while dumping chemical waste into soil or water can contaminate local ecosystems. These impacts can harm biodiversity and affect the quality of drinking water (Bouchouicha et al., 2019).

6. Social Risks

Poorly managed medical waste can also pose social risks, particularly by creating concern among the local community. Incidents of exposure to medical waste, such as needles discarded in public spaces, can lead to fear and mistrust of healthcare facilities. This can affect the reputation of hospitals and reduce public confidence in the healthcare system (Dufour et al., 2021).

7. Economic risks

The costs associated with managing medical waste can be considerable. Healthcare facilities need to invest in appropriate infrastructure, staff training and treatment processes that comply with regulations. In addition, incidents of poor waste management can lead to fines and legal action, increasing overall costs (Garrido et al., 2019).

8. Psychological risks

Medical and cleaning staff can also experience psychological risks associated with medical waste management. Stress and anxiety can be caused by fear of exposure to pathogens or hazardous chemicals. Appropriate training and psychological support can help to alleviate these concerns (Brouchou et al., 2020).

9. Risks of cross-contamination

Cross-contamination can occur when medical waste is not properly sorted. For example, if infectious waste is mixed with of recyclable waste, this can compromise the safety of recycling processes and expose the staff involved to biological risks. Appropriate training in waste sorting is therefore essential to avoid these situations (Dufour et al., 2021).

10. Legal risks

Finally, non-compliance with medical waste management regulations can expose healthcare establishments to legal risks. Laws and regulations impose strict obligations in terms of waste sorting, collection and disposal. Failure to comply with these obligations may result in sanctions, including fines and prosecution (WHO, 2021).

PATIENT AND PERSONAL SAFETY

The management of medical waste is crucial to ensuring the safety of patients and staff in University Hospitals (CHU). Inappropriate waste management can lead to serious risks, both for human health and the environment. Here are the main safety aspects of medical waste management.

1. Protection against infection

Medical waste, particularly infectious waste, represents a risk of transmission of pathogens. To minimise this risk, it is essential to put in place appropriate sorting and collection procedures. Staff must be trained to handle this waste safely, using personal protective equipment (PPE) such as gloves, masks and gowns (Brouchou et al., 2020).

2. Sharp Object Management

Sharps, such as needles and scalpels, are among the most dangerous types of medical waste. They can cause serious injuries and transmit infections. It is essential to have specific containers for collecting this waste, which must be designed to prevent punctures and cuts. Staff must also be trained in safe handling protocols (Dufour et al., 2021).

3. Training and awareness-raising

Ongoing training of staff on medical waste management practices is essential to ensure their safety. Awareness sessions should be held regularly to inform staff of the risks associated with medical waste and best management practices (Garrido et al., 2019). This training should include simulations and waste management scenarios.

4. Emergency protocols

It is important to establish emergency protocols for dealing with medical waste incidents. This includes clear procedures for dealing with sharps injuries or accidental exposure to infectious agents. Staff need to know how to respond quickly and effectively in the event of an incident (Bouchouicha et al., 2019).

5. Risk Assessment

A regular assessment of the risks associated with medical waste management must be carried out. This includes identifying critical points in the waste management process and assessing the safeguards in place. For example, temporary waste storage areas should be regularly inspected to ensure that they are secure and comply with health standards (Dufour et al., 2021).

6. Secure disposal

Proper disposal of medical waste is essential to protect the health of staff and patients. Waste must be treated using methods that comply with regulations, such as incineration or sterilisation treatment, to reduce the risk of infection and contamination. Facilities must ensure that disposal providers meet safety standards (Garrido et al., 2019).

7. Monitoring and Audit

The implementation of monitoring and audit procedures enables the effectiveness of waste management practices to be assessed. Regular audits can identify gaps in waste management and provide recommendations for improving safety. For example, checks can be carried out to ensure that staff are complying with sorting and collection protocols (Brouchou et al., 2020).

8. Internal communications

Effective communication between different departments is essential to ensure the safe management of medical waste. Information on waste management procedures should be clearly communicated to all staff, and regular meetings can be held to discuss problems and possible improvements (Dufour et al., 2021).

9. Patient Involvement

It is also important to involve patients in the safe management of waste. Healthcare facilities can educate patients on waste management practices and encourage them to report any incidents or concerns. This helps to create a safer environment for all (Garrido et al., 2019).

10. Strengthening the safety culture

Finally, it is crucial to promote a culture of safety within healthcare establishments. This involves valuing good waste management practices and recognising the importance of safety for staff and patients. A strong safety culture can reduce the risk of accidents and improve the overall management of medical waste (Bouchouicha et al., 2019).

References

Brouchou, L., Dupuy, D., & Renaud, G. (2020). Infectious Waste Management in Hospitals: Safety and Health Risks. Waste Management Research, 38(4), 345-352.

Dufour, J., Lemaire, V., & Petit, C. (2021). Sustainable Practices in Healthcare Waste Management: Challenges and Opportunities. Environment International, 150, 106-114. Garrido, M., Mendez, J., & Ruiz, A. (2019). Waste Management in Healthcare: A Review of the Literature. Journal of Environmental Management, 250, 109-115.

Bouchouicha, M., Djerrou, L., & Saidi, S. (2019). Challenges of Medical Waste Management in Algeria: A Review and Recommendations. Journal of Environmental Health Science & Engineering, 17(4), 2331-2342.

WASTE REDUCTION STRATEGIES AT SOURCE

Reducing waste at source is an essential approach to improving medical waste management in University Hospitals. By minimising waste production from the outset, the In addition to reducing costs, these facilities can improve sustainability and protect public health. Here are some effective strategies for doing so.

1. Staff awareness and training

Raising staff awareness of the importance of waste reduction is crucial. Training programmes can be set up to inform staff about best practice in waste management and encourage them to adopt responsible behaviour. For example, workshops on waste sorting and the appropriate use of resources can make staff aware of the impacts of their actions (Dufour et al., 2021).

2. Improving prescribing protocols

Optimising drug prescription protocols can reduce pharmaceutical waste. By adopting prescribing practices based on patients' real needs, UHCs can avoid overproduction of medicines and minimise waste. This includes prescribing appropriate quantities and using bulk medicines (Garrido et al., 2019).

3. Use of sustainable materials

The adoption of sustainable and reusable materials in medical procedures can also contribute to waste reduction. For example, replacing single-use medical devices with reusable alternatives, where possible and safe, can significantly reduce the amount of waste generated (Bouchouicha et al., 2019).

4. Optimisation of cleaning procedures

Cleaning and disinfection procedures can be optimised to reduce the waste generated by cleaning products. Using concentrated products and bulk cleaning systems can reduce the need for plastic packaging and cut chemical waste. In addition, efficient cleaning methods can reduce the need for disposable products (Dufour et al., 2021).

5. Sorting at source

Setting up source separation systems enables recyclable, organic and hazardous waste to be separated as soon as it is produced. This makes it easier to recycle and recover waste, thereby reducing the quantity of waste. sent to landfill. Training on waste separation should be provided to staff to ensure the effectiveness of this strategy (Garrido et al., 2019).

6. Working with Suppliers

UHCs can work with their suppliers to reduce packaging and choose products with less environmental impact. For example, negotiating supply terms that favour bulk deliveries or recyclable packaging can help reduce upstream waste (Bouchouicha et al., 2019).

7. Setting up waste reduction programmes

University hospitals can develop specific waste reduction programmes that include measurable objectives and concrete actions. For example, a programme to reduce food waste in hospital canteens may include initiatives to adjust portions served, minimise wastage and improve the management of surplus food (Dufour et al., 2021).

8. Adoption of innovative technologies

Integrating innovative technologies can make it easier to reduce waste. For example, digital management systems can help track stocks and avoid overproduction. Technological solutions can also optimise logistics processes and improve operational efficiency, helping to reduce waste production (Garrido et al., 2019).

9. Performance Evaluation and Monitoring

Evaluating and monitoring waste management performance is essential to identify areas for improvement. Performance indicators need to be defined and monitored regularly to measure the effectiveness of waste reduction strategies. This can include reports on the amount of waste produced per department and recycling rates (Bouchouicha et al., 2019).

10. Promoting a Culture of Sustainability

Finally, it is important to promote a culture of sustainability within healthcare establishments. This means encouraging all members of staff to adopt environmentally-friendly behaviour and to reflect on the impact of their actions on waste management. Initiatives such as awareness campaigns and sustainability events can reinforce this collective commitment (Dufour et al., 2021).

SUSTAINABLE PRACTICES IN HOSPITALS

Implementing sustainable practices in hospital environments is essential to reducing the environmental impact of healthcare establishments while improving the quality of care. Here is an overview of the main sustainable practices adopted in hospitals.

1. Efficient waste management

Sustainable management of medical waste includes reduction at source, effective sorting and recycling. Healthcare establishments can set up awareness programmes to train staff in appropriate sorting and disposal methods. Waste incineration and treatment should be carried out in accordance with environmental standards to minimise harmful emissions (Brouchou et al., 2020).

2. Use of environmentally-friendly materials

Adopting eco-friendly and sustainable materials for medical supplies and equipment helps to reduce the ecological footprint. This includes the use of biodegradable products, reusable medical devices and recyclable packaging. Hospitals can also choose products with a reduced environmental impact (Dufour et al., 2021).

3. Energy Efficiency

Hospitals can improve their energy efficiency by investing in energy-efficient equipment, such as LED lighting systems and low-energy heating and cooling units. The use of motion sensors and energy management systems can also reduce energy consumption in common areas (Garrido et al., 2019).

4. Renewable Energy

Integrating renewable energies into hospital infrastructures, such as solar panels or wind turbines, helps to reduce the energy consumption of our buildings. dependence on fossil fuels. This can also enable hospitals to reduce their long-term energy costs and lower their carbon footprint (Bouchouicha et al., 2019).

5. Sustainable Water Management

Efficient water management is essential to reduce wastage. Hospitals can adopt rainwater harvesting systems, install low-flow plumbing and educate staff about responsible water use. These measures can reduce drinking water consumption while ensuring an adequate supply for medical care (Dufour et al., 2021).

6. Sustainable mobility

Encouraging sustainable modes of transport for staff and patients can help reduce greenhouse gas emissions. This can include promoting carpooling, incentives for using public transport or providing bicycle parking. Hospitals can also consider partnerships with sustainable transport services (Garrido et al., 2019).

7. Sustainable Food

Hospitals can adopt sustainable food practices by favouring local, organic and seasonal produce in their canteens. This helps to reduce the carbon footprint of meals while supporting the local economy. In addition, initiatives to reduce food waste, such as redistributing surplus food, can be put in place (Boucho- uicha et al., 2019).

8. Training and awareness-raising

Ongoing staff training on sustainable practices is essential to ensure their effective implementation. Hospitals can organise workshops, seminars and awareness campaigns to encourage staff to adopt environmentally friendly behaviours (Dufour et al., 2021).

9. Integrating Sustainability into Policies

Hospitals need to integrate sustainability objectives into their management policies and strategies. This may include setting up sustainability committees, developing action plans and establishing performance indicators to monitor progress towards sustainability (Garrido et al., 2019).

10. Working with communities

Hospitals can work with local communities to promote sustainability initiatives. This can include awareness-raising projects, volunteering programmes and partnerships with environmental organisations. This collaboration fosters a collective commitment to sustainable practices (Bouchouicha et al., 2019).

WASTE MANAGEMENT INFECTIOUS WASTE

Infectious waste management is a key priority for ensuring the safety of patients, staff and the environment in University Hospital Centres (CHU). This waste, which is often contaminated with pathogens, requires specific treatment to avoid health risks. Here is an overview of best practice in infectious waste management.

1. Waste classification

It is crucial to classify infectious waste correctly as soon as it is produced. This includes biomedical waste, such as contaminated materials (needles, gloves, swabs) and waste from medical procedures. Appropriate classification makes for safer, more efficient management (Brouchou et al., 2020).

2. Staff training

Staff must receive regular training on infectious waste management protocols. Awareness sessions should cover the risks associated with infectious waste and good practice in sorting, collection and disposal. Training should include practical exercises to reinforce learning (Dufour et al., 2021).

3. Sorting at source

Sorting at source is an essential step in the management of infectious waste. Specific containers must be provided for each type of waste, with clear signage to prevent sorting errors. For example, bags can be used for infectious waste, while rigid containers are required for sharps (Garrido et al., 2019).

Image: iStock

4. Secure Collection and Transport

Infectious waste must be collected at regular intervals to avoid accumulation and the risk of contamination. Transporters must be trained to handle this waste safely, wearing appropriate personal protective equipment (PPE). Transport vehicles must be adapted to minimise the risks of leakage or spillage (Bouchouicha et al., 2019).

5. Temporary Storage

The temporary storage of infectious waste must comply with strict standards to avoid exposure. Storage areas must be clearly identified, secure and accessible only to authorised personnel. In addition, measures must be put in place to control temperature and humidity, thereby reducing the risk of decomposition and odours (Dufour et al., 2021).

6. Disposal methods

Infectious waste must be disposed of using safe methods that comply with current regulations. Commonly used techniques include : Incineration: A process for destroying waste at high temperature, reducing the volume of waste and destroying pathogens. Sterilisation: Methods such as autoclaving, which use pressurised steam to sterilise waste before disposal. Chemical decontamination: The use of chemicals to neutralise pathogens present in waste (Garrido et al., 2019).

7. Monitoring and documentation

Rigorous monitoring of infectious waste is essential to ensure that it is managed appropriately. This includes keeping records of the amount of waste generated, the disposal methods used and the dates of collection. This data is crucial for internal audits and regulatory compliance (Brouchou et al., 2020).

8. Risk Assessment

A regular assessment of the risks associated with infectious waste management must be carried out. This includes identifying critical points in the management process and assessing the safety measures in place. The results of these assessments can lead to improvements in management practices (Dufour et al., 2021).

9. Raising Patient Awareness

It is also important to raise patients' awareness of infectious waste management. Hospitals can provide information on how their waste is managed and the measures taken to ensure its safety. This helps build trust between patients and staff (Garrido et al., 2019).

10. Safety Culture

Finally, promoting a culture of safety within the hospital is essential. This means encouraging staff to report incidents of infectious waste and to take an active part in improving management practices. A strong safety culture can reduce the risk of accidents and improve overall waste management (Bouchouicha et al., 2019).

PROCEDURES AND PROTOCOLS

The management of medical waste in University Hospitals (CHU) is based on rigorous procedures and protocols designed to ensure the safety of staff, patients and the environment. First of all, it is essential to classify waste correctly as soon as it is produced. This requires clear identification of the different types of waste, such as infectious waste, chemical waste and recyclables, in order to apply appropriate management methods. Once waste has been classified, staff must be regularly trained in sorting and collection protocols. This training should include information on the risks associated with each type of waste and good handling practices. Source separation is a crucial step in the management of medical waste. Establishments must provide clearly labelled containers adapted to each type of waste, enabling effective separation as soon as it is produced. At the same time, collection protocols must be established to ensure that waste is collected at regular intervals. This prevents the accumulation of waste and reduces the risk of contamination. Waste carriers must also follow strict safety procedures, wearing appropriate personal protective equipment (PPE) to minimise the risk of exposure. Medical waste must be stored temporarily in secure, clearly identified areas accessible only to authorised personnel. Storage conditions must be monitored to prevent any degradation of the waste and to ensure safety. Disposal methods must comply with current regulations. Incineration and sterilisation are techniques commonly used to treat infectious waste, making it possible to destroy pathogens and reduce the volume of waste. Rigorous waste tracking is also necessary, with precise documentation of quantities generated, disposal methods and collection dates. Finally, it is crucial to promote a safety culture within the hospital. This means encouraging staff to report incidents relating to medical waste management and to become actively involved in improving practices. Regular audits should be carried out to assess the effectiveness of the procedures in place and identify opportunities for improvement. By integrating these elements into a systematic approach, university hospitals can ensure effective and safe management of their medical waste, thereby contributing to public health and environmental protection.

RECYCLING AND RECOVERY OF WASTE

Recycling and recovery of medical waste are essential aspects of waste management in University Hospitals (UH), with the aim of reducing environmental impact while complying with safety standards. First of all, it is crucial to identify the types of waste that can be recycled, such as plastics, paper and metals, in order to set up effective sorting systems. Sorting at source is a fundamental step, enabling recyclable waste to be separated from hazardous waste as soon as it is produced. This requires staff to be properly trained in sorting practices and to use clearly labelled containers for each type of waste. Once waste has been properly sorted, UHCs need to establish partnerships with companies specialising in waste recycling and recovery. These partnerships can include take-back schemes for recyclable materials, enabling waste to be transformed into new resources, such as recycled plastic products or paper. In addition, energy recovery, which involves recovering the energy contained in waste through processes such as controlled incineration, can also help to reduce the volume of waste while producing energy.Hospitals can also implement initiatives to reduce waste and reuse certain materials where possible. For example, sterilisable medical instruments can be reused, and uncontaminated supplies can be redistributed to other facilities or charities. At the same time, it is essential to make staff and patients aware of the importance of recycling and recovering waste. waste, running information campaigns and integrating elements of sustainability into the establishment's culture. Regular monitoring of recycling performance should be carried out to assess the effectiveness of the programmes put in place. This includes collecting data on the amount of waste recycled and recovered, as well as the savings made in terms of waste management costs. In addition, environmental audits can help to identify opportunities for improvement in waste management practices. By integrating recycling and recovery into an overall waste management strategy, UHCs can not only reduce their ecological footprint, but also contribute to a circular economy, where waste is seen as a resource rather than a burden. This approach fosters a culture of sustainability and environmental responsibility within healthcare establishments.

INNOVATIVE INITIATIVES AND TECHNOLOGIES

Innovative initiatives and technologies in medical waste management play a key role in improving the efficiency and sustainability of hospital practices. Many healthcare establishments around the world are adopting creative solutions to reduce their environmental impact and optimise their waste management. For example, some hospitals in Sweden are using pneumatic waste collection systems, enabling medical waste to be transported quickly and safely to dedicated storage areas. This reduces the risk of exposure for staff and improves the efficiency of the collection process.In Australia, advanced sterilisation technologies, such as low-temperature sterilisation systems using hydrogen peroxide, are used to treat medical waste. These systems not only decontaminate the waste, but also prepare it for recycling, thereby reducing the volume of waste sent for incineration. In addition, some healthcare establishments in Germany are adopting digital waste management platforms, which enable them to track waste volumes in real time, facilitating appropriate sorting and disposal. In the United States, the concept of 'zero waste' is becoming increasingly popular in hospitals. Institutions such as the University of Oregon Hospital have set up ambitious programmes to reduce waste at source and recycle up to 90% of their waste. This includes composting initiatives for food waste and programmes to reuse uncontaminated medical materials.In India, innovative projects are emerging to recycle medical waste, such as those that convert plastic waste into fuel. Some local companies are using pyrolysis technologies to convert plastics into oil, offering a sustainable solution for the plastic waste generated by hospitals.In South Africa, community initiatives are underway to raise awareness of medical waste management in rural clinics. Training courses are organised for healthcare staff on best sorting and disposal practices, while incorporating recycling solutions adapted to local resources.Finally, in Canada, the Toronto General Hospital has developed a programme to assess the environmental impact of its activities, which includes automated waste sorting technologies. These systems effectively separate recyclable waste from hazardous waste, reducing staff workloads and increasing recycling rates. These examples illustrate how innovation and strategic initiatives can transform medical waste management, contributing to a more sustainable future for healthcare establishments worldwide.

EDUCATION AND AWARENESS OF PERSONNEL

Educating and raising awareness among staff in healthcare establishments is essential to ensuring effective medical waste management and promoting sustainable practices. First of all, it is crucial to incorporate training programmes into the induction of new employees. These programmes should include modules on waste classification, sorting procedures and the risks associated with poor waste management. Practical sessions also reinforce learning and ensure that staff feel comfortable with the protocols in place.Secondly, ongoing awareness-raising is just as important. Regular campaigns can be put in place to remind staff of the importance of medical waste management and sustainability. This can include displays in communal areas, newsletters and dedicated team meetings. In addition, the use of visual aids, such as infographics and videos, can make it easier to understand procedures and best practice. Organising workshops and seminars on environmental and health topics, involving external experts, can also enrich staff know-how. These events provide a platform for discussing innovations in waste management and sharing experiences. Simulations of real-life scenarios, where staff have to respond to medical waste emergencies, can also enhance preparedness and responsiveness.

Image : Freepik

Another key aspect is the involvement of staff in the development of waste management policies. Teams should be encouraged to share their ideas and concerns, creating a sense of ownership and commitment to sustainable practices. Sustainability committees made up of staff from various departments can be formed to promote initiatives and improvements. In conclusion, it is essential to measure the impact of education and awareness-raising efforts. Regular surveys can be carried out to assess staff understanding and application of waste management practices. Feedback should be taken into account to adjust training programmes and ensure that they effectively meet employees' needs. By incorporating these elements, healthcare establishments can develop a strong culture of responsible waste management, thereby contributing to public health and environmental protection.

TRAINING AND INVOLVEMENT OF TEAMS

Training and team commitment are fundamental to effective medical waste management in healthcare facilities. Firstly, it is essential to develop a structured training programme that covers all aspects of waste management, including sorting, storage, disposal and associated risks. This programme should be tailored to the different levels of responsibility within the facility, ensuring that each member of staff has the appropriate knowledge to carry out their tasks safely. Interactive training sessions, including case studies and practical exercises, can reinforce learning and improve retention of information. Team commitment can be fostered by creating a culture of sustainability within the hospital. This means actively involving employees in waste management initiatives. For example, facilities can organise awareness-raising events or competitions to encourage staff to adopt better practices. Recognition of individual and collective efforts, through rewards or certification, can also reinforce this commitment. A crucial aspect of training is the ongoing assessment of skills. Regular assessments should be put in place to measure staff understanding of waste management protocols and identify areas requiring further training. Employee feedback should also be taken into account to improve training programmes and ensure that they meet the real needs of staff. Communication plays a key role in team commitment. Open communication channels, where employees can ask questions and share concerns about waste management, increase transparency and dialogue. In addition, integrating sustainability into the establishment's overall objectives and involving teams in defining these objectives can create a sense of belonging and collective responsibility. Finally, partnerships with external organisations, such as universities or waste management experts, can enrich training programmes and provide new perspectives.

References

Dufour, J., Lemaire, V., & Petit, C. (2021). Sustainable Practices in Healthcare Waste Management: Challenges and Opportunities. Environment International, 150, 106-114.

Garrido, M., Mendez, J., & Ruiz, A. (2019). Waste Management in Healthcare: A Review of the Literature. Journal of Environmental Management, 250, 109-115.

Bouchouicha, M., Djerrou, L., & Saidi, S. (2019). Challenges of Medical Waste Management in Algeria: A Review and Recommendations. Journal of Environmental Health Science & Engineering, 17(4), 2331-2342.

ROLE OF HYGIENE MANAGERS HYGIENE IN THE MANAGEMENT WASTE

The role of hygiene officers in the management of medical waste is crucial in guaranteeing the safety of patients, staff and the environment within healthcare establishments. First and foremost, these managers are responsible for developing and implementing waste management policies. They must ensure that these policies comply with current regulations and are adapted to the specific characteristics of each facility. establishment. This also involves setting up protocols for sorting, collecting and disposing of medical waste, taking into account the different types of waste and the associated risks. Secondly, hygiene managers play an essential role in staff training. They need to organise regular training sessions to make employees aware of good waste management practices, including source separation and the use of personal protective equipment (PPE). By ensuring that staff are well informed and trained, they help to minimise the risk of contamination and improve the efficiency of waste management processes. Another important aspect of their role is the monitoring and evaluation of waste management practices. They must carry out regular audits to ensure that protocols are being followed and to identify areas for improvement. In addition, they must analyse data on the volumes of waste generated and the disposal methods used, in order to optimise processes and reduce waste management costs.Hygiene managers must also work with other departments in the hospital, such as technical services and medical teams, to coordinate waste management efforts. This collaboration is essential to ensure an integrated and efficient approach, reducing the risk of errors in waste sorting and disposal.In addition, they must make all staff aware of the importance of sustainability and environmental protection in waste management. This may include initiatives to reduce waste production at source and to promote the recycling and recovery of medical waste. Finally, hygiene managers need to be up to date with new regulations and best practice in waste management. This requires ongoing training and active monitoring of innovations in the field.

BEST PRACTICES AND RESPONSIBILITIES

Best practice in medical waste management is essential to ensure the safety of patients, staff and the environment in healthcare establishments. Firstly, rigorous classification of waste as soon as it is produced is fundamental. This requires clear identification of the different types of waste, such as infectious waste, chemical waste and recyclables, in order to apply the appropriate management methods. Sorting at source must be systematic, with clearly labelled and accessible containers for each type of waste, facilitating appropriate treatment. Secondly, staff training is a crucial responsibility. Every member of the team needs to be trained in waste management protocols, including sorting, collection and disposal. Regular training sessions, including practical simulations, help to reinforce the necessary skills and ensure that staff fully understand the risks associated with poor waste management. In addition, it is important to periodically assess staff knowledge to identify any further training needs. Rigorous waste tracking is also an essential best practice. This includes accurate documentation of waste volumes produced, disposal methods used and collection dates. Managers should carry out regular audits to ensure compliance with waste management protocols and identify areas for improvement. This transparency is crucial to optimising processes and reducing the costs associated with waste management.Internal communication plays a key role in team commitment. Establishing open communication channels allows employees to express their concerns and suggest improvements to waste management. This fosters a culture of collective responsibility and involvement in sustainability initiatives. In addition, healthcare establishments need to establish partnerships with specialist waste recycling and recovery companies. These partnerships enable recyclable waste to be transformed into new products. resources and reduce the volume of waste sent for incineration. This is part of a wider approach to sustainability, where waste is seen as a resource rather than a burden. Finally, those responsible for waste management need to keep abreast of new regulations and best practice in medical waste management. Active monitoring of innovations and emerging technologies in the field is essential to continually improve waste management practices.

COLLABORATION BETWEEN THE VARIOUS DEPARTMENTS OF CHU

Collaboration between the various departments of a University Hospital Centre (CHU) is essential to ensure effective medical waste management. Each department, from medicine to logistics, plays a specific role in the waste management process, and their cooperation can improve the safety and efficiency of operations. For example, the care department must work closely with the hygiene department to ensure that staff are properly trained in waste sorting and disposal protocols. This synergy can be enhanced by cross-departmental training sessions, where experts share best practice and new regulations.Another example of effective collaboration is between the maintenance department and the waste management department. The maintenance department must ensure that waste containers are in good condition and correctly positioned to facilitate source separation. By establishing a regular communication plan between these two departments, it is possible to quickly identify equipment needs and improve the efficiency of the collection process. In addition, regular meetings can be set up to assess compliance with waste management practices and suggest improvements.The pharmacy department also plays a crucial role in waste management, particularly when it comes to unused or out-of-date medicines. Close collaboration between the medical teams and the pharmacy ensures that this waste is correctly identified and disposed of in accordance with regulations. For example, some hospitals have set up Drug return programmes, where patients can return unused medicines, reducing the risk of contamination and wastage. Patient awareness is another area where cross-departmental collaboration can be beneficial. Care teams, in partnership with the communications department, can create explanatory brochures and posters on medical waste management, raising awareness among patients and their families of the importance of waste management. This helps to create a safer and more environmentally aware healthcare environment. Lastly, research projects and continuous improvement initiatives may be carried out in collaboration with universities or external organisations, enabling UHC departments to benefit from the latest innovations in waste management. For example, a university hospital in France worked with a university to develop a digital waste tracking system, optimising management and reducing sorting errors.

THE IMPORTANCE OF INTERDISCIPLINARY WORK

Interdisciplinary work is crucial in the management of medical waste within University Hospitals (UH), as it enables the integration of diverse expertise and skills to improve the efficiency and safety of processes. Each discipline, whether medical, administrative, technical or environmental, brings a unique perspective that enriches overall waste management.

For example, medical teams, in collaboration with hygiene managers, can better identify the types of waste generated during patient care, facilitating appropriate classification and sorting at source.Another important aspect is the collaboration between care staff and technical services. Care teams need to work with maintenance departments to ensure that waste containers are correctly positioned and adapted to the needs of the different departments.

This includes establishing protocols for the cleaning and maintenance of waste storage areas, thus ensuring a safe environment for staff and patients. patients. For example, a university hospital in Italy set up regular meetings between nursing staff and the maintenance department to discuss the improvements needed, which led to a significant reduction in sorting errors.

The role of waste management teams is also essential in an interdisciplinary setting.

These teams need to work with the pharmacy department to properly manage pharmaceutical waste, such as expired or unused medicines. By working together, they can develop clear protocols for the return and disposal of this waste, reducing the risk of contamination and waste. In addition, the involvement of communications departments is vital in raising awareness among staff and patients of the importance of waste management. Working with medical teams, the communications department can create educational materials that explain sorting practices, the risks associated with poor waste management and the benefits of a sustainable approach.

For example, informative brochures and posters can be displayed in waiting rooms and communal areas, raising awareness and getting everyone involved. Finally, interdisciplinary work encourages innovation and continuous improvement.

By bringing together experts from different fields, UHCs can identify new technologies and best practices for waste management. For example, a university hospital in Australia formed a working group made up of medical staff, environmental technicians and waste management experts to look for innovative solutions for recycling and recovering waste, leading to significant results in terms of sustainability. In summary, the importance of interdisciplinary work in medical waste management lies in its ability to create a coherent and integrated approach. This collaboration enriches knowledge, improves process efficiency and enhances safety within healthcare establishments, thereby contributing to a more sustainable and responsible healthcare environment.

WASTE MANAGEMENT PHARMACEUTICALS

Pharmaceutical waste management is a crucial aspect of medical waste management in University Hospitals (UHCs) and requires rigorous protocols to ensure the safety of staff, patients and the environment. Firstly, it is essential to classify pharmaceutical waste as soon as it is produced. This includes out-of-date, unused or contaminated medicines, which must be separated from other types of waste to avoid cross-contamination. Each facility must have clearly labelled and appropriate containers for storing this waste, to ensure that it is disposed of properly.Secondly, staff training is essential. Medical teams, as well as pharmacy staff, need to be trained in pharmaceutical waste management procedures, including the risks associated with mishandling this waste. Regular training sessions, accompanied by visual aids and practical simulations, help to reinforce staff understanding and application of the protocols. Another major challenge is the safe disposal of pharmaceutical waste. University hospitals need to work with specialist companies to ensure that this waste is treated in accordance with current regulations. For example, some establishments may use dedicated collection services that take care of pharmaceutical waste, ensuring that it is disposed of in a safe and environmentally-friendly manner. In some countries, drug return programmes are also in place, allowing patients to bring back unused medicines for proper disposal. Raising patients' awareness of pharmaceutical waste management is also important. Hospitals can develop information campaigns to explain to patients how to properly dispose of expired or unused medicines, providing clear instructions on available collection points. This helps to reduce the risk of contamination and promote responsible healthcare practices. In addition, the monitoring and evaluation of pharmaceutical waste management practices are essential for the continuous improvement of processes. Regular audits should be carried out to check compliance with the protocols in place and identify areas requiring adjustment. This also makes it possible to analyse the volumes of waste generated and identify opportunities to reduce waste. Finally, interdisciplinary collaboration is a key element in successful pharmaceutical waste management. Medical teams, pharmacists, hygiene staff and technical departments need to work together to develop effective protocols that are tailored to the needs of the facility. For example, interdepartmental meetings can be organised to discuss the challenges encountered and share innovative solutions.

RISKS AND METHODS OF TREATMENT

The management of medical waste, particularly infectious and pharmaceutical waste, presents significant risks that require appropriate treatment methods to ensure the safety of patients, staff and the environment. Firstly, medical waste can contain pathogens, hazardous chemicals and out-of-date medicines, posing risks of contamination, infection and pollution. Inappropriate exposure to this waste can lead to nosocomial infections, allergic reactions or long-term health problems for healthcare staff and patients. To minimise these risks, it is essential to implement rigorous sorting protocols. Waste must be classified at source, with clearly identified containers for infectious waste, chemical waste and medicines. This separation allows each type of waste to be treated by the most appropriate method, reducing the potential for cross-contamination. Treatment methods for medical waste vary according to its nature. Incineration is one of the most commonly used methods for eliminating infectious waste. This process destroys pathogens at high temperature, reducing the volume of waste and minimising the environmental impact. However, it requires specific facilities and rigorous monitoring to avoid harmful emissions. Autoclave sterilisation is another effective method, particularly suited to infectious waste. This process uses pressurised steam to destroy pathogens, allowing the waste to be disposed of by conventional means. Many hospitals are adopting this method because of its ability to treat waste effectively while complying with environmental standards. For chemical and pharmaceutical waste, specific decontamination methods must be applied. These may include chemical neutralisation processes, where chemicals are added to neutralise the toxicity of the waste before disposal. In addition, some facilities use advanced treatment technologies, such as pyrolysis, which converts waste into energy, contributing to a more sustainable approach. Staff training is also essential to ensure that treatment methods are applied correctly. Teams need to be made aware of the risks associated with different types of waste and the treatment protocols to be followed. Simulations and practical exercises can reinforce this training, ensuring that staff are well prepared to deal with emergency situations. Finally, the monitoring and evaluation of the treatment methods in place are crucial to the continuous improvement of practices. Regular audits identify weak points and suggest improvements, ensuring that protocols remain compliant with regulations and best practice in waste management.

INNOVATIVE TECHNOLOGIES FOR WASTE TREATMENT

Innovative technologies for the treatment of medical waste play a crucial role in improving waste management in University Hospitals (UH). These solutions not only reduce health risks, but also minimise environmental impact. Microwave sterilisation, for example, is a method that is gaining in popularity. This technology uses electromagnetic waves to heat and destroy the pathogens present in waste. infectious diseases, enabling effective treatment while maintaining a reduced environmental footprint. Another example is the plasma decontamination technique. This method uses ionised plasma to eliminate pathogens and toxic substances from medical waste. Facilities in Europe and the United States have begun to adopt this technology, which offers a less polluting alternative to traditional incineration methods, while being capable of treating a variety of waste, including chemical waste. In addition, the use of pyrolysis technologies to treat medical plastic waste is increasingly being considered. This thermochemical process breaks down materials at high temperatures in the absence of oxygen, producing gases and oils that can be used as energy sources. For example, some facilities in Asia have implemented this technology to convert plastic waste into fuel, contributing to a circular economy. Digitisation and the use of intelligent sensors also represent a significant advance in waste management. These systems make it possible to track the volume of waste produced in real time and monitor its treatment. For example, some hospitals in Canada have implemented waste management software that incorporates sensors to optimise waste collection and sorting, reducing errors and improving operational efficiency. Other innovations include the use of robots to sort waste. These automated systems can recognise different types of waste and sort them according to their nature, reducing staff workload and minimising the risk of exposure. Facilities in Europe have already experimented with these technologies with promising results, increasing sorting accuracy and staff safety. Finally, research projects are underway into the recovery of medical waste, in particular the development of biotechnologies to transform organic waste into compost or biogas. For example, some hospitals in Scandinavia are exploring the use of biodigesters to treat medical waste. and organic waste, thereby contributing to sustainability and reducing the amount of waste sent to landfill. These solutions enable waste to be processed more effectively, while minimising environmental impact and guaranteeing the safety of patients and staff.

CONTEMPORARY EQUIPMENT AND SOLUTIONS

Today's medical waste processing equipment and solutions are essential for improving the efficiency and safety of operations in University Hospitals (UH). Among these solutions, steam treatment systems, such as autoclaves, are widely used to sterilise infectious waste. This equipment works by using pressurised steam to destroy pathogens, enabling waste to be disposed of safely and prepared for disposal at appropriate sites. Another contemporary solution is the use of intelligent waste containers. These devices are equipped with sensors that monitor the fill level and alert staff when the container is almost full. This optimises waste collection, reduces the risk of overflow and improves the efficiency of waste management operations. Hospitals in France, for example, have begun to deploy these technologies to improve their collection logistics. The microwave treatment solutions mentioned above also represent a significant advance. This equipment effectively treats medical waste by heating it rapidly, thereby reducing the volume of waste while ensuring decontamination. Facilities in Switzerland have adopted this technology, seeing a significant reduction in the costs associated with waste disposal. Low-emission incineration systems are another contemporary solution. These modern incinerators are designed to minimise pollutant emissions into the environment, while ensuring efficient destruction of medical waste. For example, hospitals in the UK have invested in new-generation incinerators that meet strict environmental standards, reducing their ecological footprint. In addition, the development of biodigesters to treat organic waste, such as food scraps or waste from care services, is booming. This equipment converts organic waste into biogas, which can be used as a source of renewable energy. Pilot projects in Scandinavia show that this approach can not only reduce waste, but also contribute to the energy sustainability of establishments. Finally, waste management software, which incorporates waste analysis and tracking tools, is being used more and more. These solutions enable facilities to monitor their waste flows, optimise their sorting and disposal processes, and improve their sustainability reporting. Hospitals in Australia have implemented these systems to analyse waste data, identifying opportunities for reduction and continuous improvement. These innovations not only help to protect public health, but also reduce the environmental impact of hospital activities.

MONITORING AND EVALUATION OF WASTE MANAGEMENT

The monitoring and evaluation of medical waste management is essential to ensure efficient processes, regulatory compliance and continuous improvement of practices within University Hospitals (UHCs). Firstly, it is crucial to put in place clear performance indicators to measure the effectiveness of waste management systems. These indicators can include measures such as the volumes of waste generated, sorting and recycling rates, as well as the costs associated with waste disposal. For example, a hospital can set up a monthly reporting system to analyse this data and identify trends. The use of waste management software is also a contemporary practice that facilitates monitoring. These systems enable waste data to be centralised, waste flows to be tracked in real time and waste to be recycled. automate reporting. For example, hospitals in Canada have adopted digital platforms to monitor the entire waste management process, making it easier to analyse and optimise practices.

Image : Freepik

Regular audits play a key role in assessing waste management practices. These audits should be carried out by internal or external teams to ensure objectivity and transparency. They enable compliance with established protocols to be assessed, shortcomings in management to be identified and improvements to be recommended. For example, a university hospital in France set up a six-monthly audit programme to assess the effectiveness of its waste management practices, which led to significant improvements in waste sorting and reduction. Staff awareness and ongoing training are also essential for effective monitoring. Regular training sessions remind employees of good waste management practices and assess their understanding of the protocols. Satisfaction questionnaires or evaluations can be used to measure the impact of these training sessions and adjust the programmes as required. Finally, collecting feedback from staff is a valuable practice. Encouraging staff to report problems encountered in waste management helps to identify opportunities for improvement. For example, some hospitals set up suggestion boxes or monthly meetings to gather ideas and concerns from employees in the field.

CASE STUDIES OF UNIVERSITY HOSPITALS IN EUROPE

Here are a few case studies from University Hospitals in Europe that illustrate innovative and effective practices in medical waste management:

1. Bordeaux University Hospital, France

Bordeaux University Hospital has set up an ambitious waste reduction programme, which includes integrating waste management into its sustainable development strategy. This programme includes:
-Ongoing training: Regular sessions to raise staff awareness of the importance of waste sorting.
Audit and monitoring: Half-yearly assessments to measure progress in waste reduction.
-Recycling: Implementation of recycling circuits for plastics, paper and metals, leading to a significant reduction in the volume of waste sent for incineration.

2. Copenhagen University Hospital, Denmark

This hospital has adopted a systematic approach to managing its medical waste, with impressive results:

-Treatment technology: Use of a modern incinerator that complies with strict environmental standards and reduces pollutant emissions.
-Interdisciplinary collaboration: medical, technical and administrative teams work together to optimise the sorting and disposal processes.
-Raising patient awareness: Information campaigns to encourage patients to return unused medicines, thereby reducing pharmaceutical waste.

3. Charité Hospital, Berlin, Germany

This hospital has implemented an exemplary waste management programme, focused on innovation :

-Automated sorting systems : Adoption of automated sorting technologies that improve the efficiency of the waste management process.
-Biowaste: Implementation of a composting system for organic waste, transforming food waste into compost used in the hospital's green spaces.
-Sustainability report: Annual publication of reports on waste management, enabling progress to be monitored and best practice to be shared.

4. Lausanne University Hospital, Switzerland

Lausanne University Hospital has developed an integrated waste management programme that illustrates the importance of sustainability:

-Training and awareness: Training programmes for all employees on waste management practices.
Continuous assessment: Regular audits to measure compliance with waste management protocols, with built-in feedback to improve processes.
-External partnerships: Working with local companies to recycle plastic waste, reducing the amount sent for incineration.

5. Göteborg University Hospital, Sweden

This hospital has adopted a proactive approach to the management of its medical waste:
Low-emission incineration: Use of low-emission incinerators,
respecting the strictest environmental standards.
Digital tracking: Implementation of a digital platform for tracking waste, enabling real-time management and better decision-making.
-Waste reduction initiatives: Programmes to reduce packaging and promote the use of sustainable materials.

Conclusion

These case studies show how Europe's university hospitals are adopting innovative and collaborative practices to improve medical waste management. These establishments are contributing to a more sustainable and responsible healthcare environment. These initiatives can serve as models for other establishments seeking to optimise their waste management.

SUCCESS STORIES AT USA

Here are a few success stories on medical waste management in the United States, illustrating effective and innovative practices implemented by hospitals and health systems:

1. Mayo Clinic, Minnesota

The Mayo Clinic is recognised for its sustainability and waste management initiatives:

-Waste reduction programme: The clinic has a programme in place that aims to reduce waste by 50% by 2030. This includes efforts to reduce packaging and improve recycling.
-Staff training: Regular training sessions are organised to raise staff awareness of waste management, with particular emphasis on sorting at source.
-Monitoring progress: Use of monitoring systems to measure volumes of waste generated and assess progress against reduction targets.

2. Cleveland Clinic, Ohio

The Cleveland Clinic has adopted best practices in medical waste management:
Low-emission incineration: The facility uses incinerators that comply with strict environmental standards, reducing pollutant emissions.
-Partnerships with local businesses: Working with specialist companies to recycle plastic and metal waste, contributing to a circular economy.
-Sustainability reporting: Annual reporting on sustainability and waste management performance, ensuring transparency and accountability.

3. Kaiser Permanente, California

Kaiser Permanente is an integrated healthcare system that focuses on sustainability:

-Waste reduction initiatives: Implementation of waste reduction programmes across all sites, including efforts to reduce food waste and improve recycling.
-Employee awareness: Ongoing training for staff on best practice in waste management, with incentives for those who actively participate in sustainability programmes.
-Monitoring technologies: Use of digital technologies to monitor waste production and optimise management processes.

4. Johns Hopkins Hospital, Maryland

The Johns Hopkins Hospital is an example of effective medical waste management:

Sorting and disposal strategy: Implementation of a rigorous sorting system for medical waste, with clearly labelled containers and well-defined collection procedures.
-Regular audits: Internal audits are carried out to assess compliance with waste management protocols and identify areas for improvement.
-Community involvement: Raising community awareness of medical waste management, including return programmes for unused medicines.

5. Children's Hospital Colorado

This paediatric hospital has introduced innovative waste management practices:
-Biodegradability: Adoption of biodegradable products in food and care services, thereby reducing plastic waste.
-Composting programme: Implementation of a composting programme for organic waste, contributing to sustainability and reducing the amount of waste sent to landfill.
-Impact assessments: Regular assessments of the environmental impacts of waste management practices to guide future strategic decisions.

Conclusion

These successful models from the United States show how hospitals and healthcare systems are adopting innovative and effective practices to manage medical waste. These initiatives can serve as inspiring examples for other institutions seeking to improve their waste management.

ENVIRONMENTAL IMPACT OF HOSPITAL WASTE

The environmental impact of hospital waste is a major issue that requires particular attention in the management of healthcare establishments. This waste, which includes infectious, chemical and pharmaceutical products, can have harmful consequences for the environment if it is not treated properly. For example, the incineration of medical waste can release pollutants into the air, affecting air quality and contributing to public health problems. In the United States, studies have shown that hospitals that do not comply with incineration standards can emit dioxins, carcinogenic substances, into the atmosphere. In Europe, hospital waste management is governed by strict regulations, but challenges persist. In France, for example, hospitals have been criticised for failing to sort waste properly, leading to increased volumes of waste being sent for incineration and, as a result, greater greenhouse gas emissions. Poor waste management can also lead to soil and groundwater contamination, especially when chemical waste is disposed of incorrectly. This poses a threat to local biodiversity and can have long-term repercussions on ecosystems. Plastic waste is another source of environmental concern in healthcare establishments. With the increase in single-use medical products, such as gloves and syringes, the amount of plastic generated by hospitals has risen considerably. In the US, it has been estimated that hospitals generate around 2 million tonnes of plastic waste per year, much of which ends up in landfill, contributing to the global plastic waste crisis. In Europe, initiatives are underway to reduce the use of single-use plastics in healthcare establishments, but the challenge remains immense.

Image : Freepik

Improper treatment of medical waste can also have an impact on human health. For example, as a result of inappropriate management, pathogens present in infectious waste can contaminate the environment and pose risks to surrounding communities. Cases of contamination of aquifers by poorly managed hospital waste have been reported in Italy, where toxic substances have been found in drinking water, endangering the health of local residents.Recovering medical waste by recycling and reusing it is a sustainable alternative. Hospitals in Sweden have set up recycling programmes for medical plastics, turning waste into new resources and reducing the environmental impact of their operations. waste. Similarly, initiatives in the United States, such as those launched by the Cleveland Clinic, aim to integrate sustainability practices into waste management, thereby reducing the volumes of waste sent for incineration and improving the carbon footprint of facilities. Educating and raising awareness among medical staff about the environmental impact of hospital waste is essential to improving management practices. Training programmes can help to make staff aware of the consequences of poor waste management and encourage them to adopt more responsible behaviour. For example, the Mayo Clinic has set up training sessions for its staff to encourage them to sort and reduce waste. Collaboration between healthcare establishments, governments and non-governmental organisations is also crucial to reducing the environmental impact of hospital waste. In Europe, projects such as the Green Hospital programme aim to create synergies between different stakeholders to share best practice and develop innovative solutions. In the United States, similar initiatives encourage hospitals to work together to reduce their collective environmental footprint. In conclusion, the environmental impact of hospital waste is a complex issue that requires a multi-dimensional approach. By adopting sustainable management practices, raising staff awareness and working with other stakeholders, healthcare establishments can significantly reduce their impact on the environment. This is essential not only for the health of surrounding communities, but also for the protection of the environment and the long-term sustainability of healthcare systems.

ANALYSIS OF CONSEQUENCES

Analysis of the ecological consequences of hospital waste reveals a significant impact on the environment, requiring greater attention in waste management practices. Firstly, medical waste, particularly that containing toxic or infectious substances, can contaminate soil and groundwater. For example, studies carried out in Italy have shown that poorly managed chemical waste has penetrated aquifers, leading to alarming levels of contaminants in drinking water. This situation poses direct risks to human health and local biodiversity. Secondly, the incineration of medical waste, although commonly used to destroy pathogens, produces harmful emissions that affect air quality. In the United States, hospitals that do not comply with emission standards can release dioxins and other pollutants into the atmosphere, contributing to public health problems such as respiratory diseases and cancer. In Europe, initiatives to modernise incinerators have been put in place to reduce these emissions, but challenges remain. Another worrying aspect is the accumulation of plastic waste. Healthcare establishments generate a colossal amount of plastic waste, mainly due to the use of single-use products. In the United States, it is estimated that hospitals produce around 2 million tonnes of plastic waste a year. Most of this plastic ends up in landfill, where it can take hundreds of years to decompose, contributing to the global plastic waste crisis. In addition, poor management of biological waste can lead to the proliferation of pathogens in the environment. Infectious waste, if not treated properly, can release bacteria and viruses into the soil and water, increasing the risk of infections in surrounding communities. Cases of contamination of rivers by medical waste have been reported in Europe, underlining the importance of rigorous management. Ecological consequences are not limited to direct impacts on the environment. They also include long-term effects on the health of ecosystems. Soil and water contamination can affect flora and fauna, disrupting natural habitats and threatening local species. For example, chemicals from hospital waste can accumulate in the food chain, affecting not only aquatic organisms, but also land animals and, ultimately, humans. In addition, inappropriate treatment of medical waste can also lead to greenhouse gas emissions. Waste incineration, in particular, produces carbon dioxide and other harmful gases that contribute to climate change. Studies in the United States show that hospitals, as emitting entities, must take steps to reduce their carbon footprint through more sustainable waste management. Raising awareness and training healthcare professionals are essential to mitigate these environmental consequences. By training staff in best waste management practices, facilities can reduce errors and improve waste sorting, collection and disposal. Awareness-raising programmes, such as those set up by the Mayo Clinic, can have a significant impact on reducing the amount of waste produced. Finally, collaboration between healthcare establishments, governments and environmental organisations is crucial to developing innovative and sustainable solutions. Initiatives in Europe, such as the Green Hospital programme, encourage the sharing of best practice and innovation in waste management. In the United States, similar partnerships aim to reduce the collective ecological impact of hospitals and promote sustainable practices. In conclusion, an analysis of the ecological consequences of hospital waste highlights the major challenges facing healthcare establishments. A proactive approach, incorporating sustainable management practices, staff awareness-raising and collaboration with various stakeholders, is essential to minimise these impacts and protect the environment.

FUTURE PROSPECTS FOR WASTE MANAGEMENT AT CHU

The future outlook for waste management in University Hospitals (UH) looks promising, with a trend towards more sustainable and innovative practices. Firstly, the integration of advanced technologies into the waste management process is set to become the norm. Automated sorting systems, for example, will improve the efficiency of sorting at the source, thereby reducing errors and increasing recycling rates. Solutions such as artificial intelligence could also be used to optimise waste management, by analysing flows and predicting needs according to hospital activities. Secondly, awareness of sustainability will continue to grow within healthcare establishments. Teaching staff and patients about the importance of waste management will be an increasing focus for UHCs. Educational programmes and awareness campaigns should be developed to encourage responsible behaviour, including waste sorting and reduction. For example, initiatives at institutional level could encourage patients to return unused medicines, thereby limiting pharmaceutical waste. Inter-institutional collaboration should also be strengthened. University hospitals could join forces with other healthcare institutions, private companies and non-governmental organisations to share best practice and innovate in waste management. Regional collaboration networks could facilitate the exchange of ideas and resources, thereby improving operational efficiency and reducing waste management costs. Regulation and public policy will play a crucial role in the evolution of waste management. Governments should introduce stricter standards for the treatment of medical waste, thereby encouraging institutions to adopt more sustainable practices. For example, tax incentives could be offered to hospitals that invest in environmentally-friendly waste treatment technologies. This could also include regulations to reduce the use of single-use plastics in healthcare facilities. Greater attention will also be paid to circularity in waste management. UHCs will seek to integrate material recovery and reuse practices, transforming waste into resources. Advanced recycling initiatives, such as transforming plastic waste into new raw materials, could become commonplace. For example, Some hospitals could adopt biodegradation systems to treat organic waste and reduce their overall volume of waste. In addition, the monitoring and evaluation of waste management practices will become more sophisticated. The use of real-time data will enable facilities to monitor their performance and identify areas for improvement. Analysis of the data collected will also help establish benchmarks and compare performance between different facilities, fostering a culture of continuous improvement. Research and development will also play a key role in waste management innovation.

University hospitals could collaborate with universities and research institutes to develop new technologies and waste treatment methods, thereby strengthening their ability to manage medical waste effectively. Finally, growing environmental concerns and the urgency of the climate crisis will inevitably influence waste management practices in UHCs. Healthcare establishments will need to respond to these challenges by adopting greener practices and contributing to the overall sustainability of their communities. This includes not only waste management, but also thinking more broadly about their ecological footprint and their contribution to public health.

ROLE OF EXTERNAL PARTNERS

The role of external partners in waste management within University Hospitals (UH) is crucial to improving sustainability practices and enhancing operational efficiency. Firstly, companies specialising in the treatment of medical waste play an essential role by offering solutions tailored to the specific characteristics of the waste generated by healthcare establishments. For example, companies specialising in waste incineration or sterilisation can ensure that waste is treated in accordance with current regulations, thereby reducing the risks to public health and the environment. Secondly, non-governmental organisations (NGOs) and environmental associations can provide valuable expertise in awareness-raising and education. These partners can work with UHCs to develop waste management awareness programmes aimed at encouraging staff and patients to adopt responsible behaviour. For example, initiatives that raise awareness of plastic waste reduction or the need to return unused medicines can have a significant impact on the amount of waste generated. Public authorities also play an essential role in hospital waste management. By introducing strict regulations and environmental standards, they can encourage hospitals to adopt more sustainable waste management practices. For example, subsidies or tax incentives can be offered to establishments that invest in environmentally-friendly waste treatment technologies, thereby encouraging a transition to more sustainable practices. Universities and research institutes also contribute to innovation in medical waste management. By collaborating with university hospitals, these institutions can conduct research into new waste treatment, recycling and materials recovery technologies. For example, research projects focusing on bioenergy could make it possible to transform organic waste into renewable energy sources, thereby reducing dependence on fossil fuels. Logistics and transport companies play a key role in the efficient management of waste flows. By optimising collection routes and improving the management of waste stocks, these external partners can help to reduce the costs and carbon footprint of waste management operations. For example, the use of waste management software enables healthcare establishments to monitor waste volumes in real time and optimise collections. Community initiatives can also strengthen hospital waste management. By engaging with local communities, UHCs can raise public awareness of the importance of waste management and encourage collective action, such as clean-up days or drug return programmes. This not only promotes better waste management, but also a positive relationship between the hospital and the community. International collaboration can also provide valuable insights. By sharing best practice with healthcare establishments in other countries, UHCs can learn about new approaches and technologies for waste management. For example, international conferences and forums on sustainability in healthcare allow professionals to share experiences and adopt proven innovations. Finally, recycling and waste recovery companies can help UHCs transform their waste into resources. By partnering with these companies, healthcare institutions can not only reduce the amount of waste sent to landfill, but also contribute to a circular economy. For example, initiatives to recycle medical plastics into new raw materials can turn a problem into an opportunity. External partners play a fundamental role in waste management at UHCs. Their expertise, resources and commitment to sustainability can significantly improve waste management practices, contributing to a safer and more sustainable healthcare environment. By collaborating with these players, healthcare establishments can not only optimise their waste management, but also strengthen their positive impact on public health and the environment.

COMMUNITY COMMITMENT TO WASTE MANAGEMENT

Community involvement in waste management at University Hospitals (UHCs) is an essential element in promoting sustainable practices and strengthening the relationship between hospitals and their communities. Firstly, raising community awareness of the importance of medical waste management can help to reduce the amount of waste generated. Awareness campaigns, organised in collaboration with schools, local organisations and community groups, can inform the public about the dangers associated with medical waste and encourage responsible behaviour, such as returning unused medicines.

Image : Freepik

In addition, UHCs can organise community events to promote waste reduction and recycling. These events, such as clean-up days or waste-sorting workshops, allow community members to get involved in concrete actions. For example, a hospital in Germany organised awareness days where volunteers helped to sort waste, while learning about medical waste management practices. These initiatives strengthen the sense of belonging and encourage a culture of sustainability within the community. Partnerships with local organisations can also enrich waste management efforts. By working with NGOs and environmental groups, UHCs can benefit from the expertise of these organisations to develop awareness and training programmes. For example, NGOs can help design educational programmes on the dangers of medical waste and on best practices. for their disposal. This creates a synergy between healthcare establishments and the community, strengthening the collective capacity to manage waste. Community involvement can also take the form of waste collection initiatives. University hospitals can organise medicine return programmes, where patients and community members can return unused or out-of-date medicines for safe disposal. These programmes not only reduce the risk of pollution, but also encourage active community participation in waste management. For example, several hospitals in the United States have set up collection points for medicines, attracting the involvement of residents. Healthcare establishments can also integrate environmental concerns into their relations with the community. By organising public forums or meetings, university hospitals can listen to citizens' concerns about waste management and answer their questions. This transparency helps build trust and shows that hospitals take their responsibility for the environment and public health seriously. In addition, volunteering initiatives can strengthen community involvement. UHCs can encourage their employees to take part in clean-up activities in the surrounding neighbourhoods or in educational programmes in schools. These actions demonstrate a commitment to the community and strengthen the ties between hospital staff and residents. Hospitals in Canada, for example, have encouraged their employees to organise workshops in schools on waste management, thereby raising awareness of responsible practices among young people. Community involvement can also foster innovation in waste management. The ideas and concerns expressed by community members can help develop creative solutions tailored to local needs. For example, community groups can propose composting initiatives for organic waste generated by hospitals, promoting a sustainable approach and reducing waste sent to landfill. Finally, community involvement in waste management helps to reinforce the social responsibility of university hospitals. By showing that they care about the environment and the health of the community, healthcare establishments can improve their image and win public support. It can also encourage more collaboration between hospitals and communities, paving the way for future initiatives that promote sustainability and public health.

CONCLUSION AND CALL TO ACTION

In conclusion, waste management in University Hospitals is a crucial issue that impacts not only public health, but also the environment. The challenges posed by medical waste, whether infectious, chemical or plastic, require an integrated and collaborative approach. The commitment of all stakeholders (healthcare establishments, communities, external partners, etc.) is essential to developing sustainable and effective solutions.It is imperative that UHCs implement waste management practices that comply with environmental standards while protecting the health of patients and employees. This includes adopting innovative technologies, raising awareness among staff and patients, and strengthening collaboration with external partners. Community initiatives also play a key role in mobilising citizens around waste management, fostering behavioural change and a culture of sustainability. In this respect, we are calling for collective action. Hospitals must step up their training and awareness programmes, and governments must establish stricter regulations for the management of medical waste. Communities must become actively involved in initiatives to raise awareness and reduce waste. Every player has a role to play in this dynamic. Together, we can turn the challenges of waste management into opportunities to create more sustainable and responsible healthcare systems. By adopting a proactive approach, we can not only protect our It's not just about protecting the environment, it's also about ensuring a healthy future for generations to come. It's time to act, to work together and to make the management of medical waste a priority in our healthcare establishments.

FINAL THOUGHTS AND PROSPECTS

The final thoughts on waste management in University Hospitals (UHCs) underline the importance of a systemic and collaborative approach. The complexity of medical waste, which includes infectious, chemical and plastic materials, requires not only rigorous management strategies, but also a heightened awareness of environmental and public health issues. In this respect, it is essential that UHC adopt sustainable practices that minimise their ecological footprint while guaranteeing the safety of patients and staff. A key prospect for the future is the integration of digital technologies into waste management. Real-time tracking systems, based on waste management software, can enable hospitals to better control their waste streams, optimise sorting and improve collection efficiency.

Examples such as the Cleveland Clinic's waste management system show how digitisation can transform waste management by providing accurate data and facilitating decision-making. In addition, education and awareness-raising play a crucial role in changing behaviour within healthcare establishments. Ongoing training programmes for staff and awareness-raising initiatives for patients can have a significant impact on waste reduction.

For example, the Mayo Clinic has introduced training sessions on waste management, which has led to an increase in recycling rates and a reduction in plastic waste. Partnerships with external stakeholders, such as NGOs, recycling companies and public authorities, can also strengthen UHC's waste management efforts. By working with organisations such as "Health Care Without Harm", which aims to reduce the environmental impact of healthcare establishments, UHCs can benefit from best practice and invaluable expertise. In addition, initiatives such as the "Green Hospital" programme in Europe encourage the exchange of knowledge between establishments to develop innovative solutions. Future prospects for waste management in UHC also include the adoption of circular business models.

This involves not only reducing waste at source, but also reusing and recycling materials where possible. Institutions such as the Charité Hospital in Berlin have already begun to implement advanced recycling programmes, transforming plastic waste into new raw materials. Finally, it is imperative that public policies support these initiatives. Stricter regulations and financial incentives for sustainable practices can encourage UHC to invest in environmentally-friendly solutions.

Examples from France, where subsidies are provided for the modernisation of waste treatment systems, illustrate how government support can act as a catalyst for change. Waste management in university hospitals is an evolving field, where sustainable practices, increased awareness and effective collaboration can turn challenges into opportunities. It's time to act collectively to build a sustainable future in healthcare.

BIBLIOGRAPHICAL SOURCES AND REFERENCES

World Health Organization (WHO). (2021). Health Care Waste Management: A Resource for Health Care Facilities. Geneva: WHO Press.

Brouchou, L., Dupuy, D., & Renaud, G. (2020). Infectious Waste Management in Hospitals: Safety and Health Risks. Waste Management Research, 38(4), 345-352.

Dufour, J., Lemaire, V., & Petit, C. (2021). Sustainable Practices in Healthcare Waste Management: Challenges and Opportunities. Environment International, 150, 106- 114.

Garrido, M., Mendez, J., & Ruiz, A. (2019). Waste Management in Healthcare: A Review of the Literature. Journal of Environmental Management, 250, 109-115.

Dufour, J., Lemaire, V., & Petit, C. (2021). Sustainable Practices in Healthcare Waste Management: Challenges and Opportunities. Environment International, 150, 106- 114.

Brouchou, L., Dupuy, D., & Renaud, G. (2020). Infectious Waste Management in Hospitals: Safety and Health Risks. Waste Management Research, 38(4), 345-352.

Garrido, M., Mendez, J., & Ruiz, A. (2019). Waste Management in Healthcare: A Review of the Literature. Journal of Environmental Management, 250, 109-115.

World Health Organization (WHO). (2021). Health Care Waste Management: A Resource for Health Care Facilities. Geneva: WHO Press.

European Commission. (2008). Directive 2008/98/EC on waste.

European Commission. (2011). Directive 2011/97/EU on waste from healthcare activities presenting an infectious risk.

ISO. (2015). ISO 14001:2015 - Environmental management systems.

Dufour, J., Lemaire, V., & Petit, C. (2021). Sustainable Practices in Healthcare Waste Management: Challenges and Opportunities. Environment International, 150, 106- 114.

Brouchou, L., Dupuy, D., & Renaud, G. (2020). Infectious Waste Management in Hospitals: Safety and Health Risks. Waste Management Research, 38(4), 345-352.

European Commission. (1989). Directive 89/391/EEC on the health and safety of workers.

European Commission. (2012). Directive 2012/19/EU on waste electrical and electronic equipment (WEEE).

Brouchou, L., Dupuy, D., & Renaud, G. (2020). Infectious Waste Management

in Hospitals: Safety and Health Risks. Waste Management Research, 38(4), 345-352.
Dufour, J., Lemaire, V., & Petit, C. (2021). Sustainable Practices in Healthcare Waste Management: Challenges and Opportunities. Environment International, 150, 106- 114.
Garrido, M., Mendez, J., & Ruiz, A. (2019). Waste Management in Healthcare: A Review of the Literature. Journal of Environmental Management, 250, 109-115.
Bouchouicha, M., Djerrou, L., & Saidi, S. (2019). Challenges of Medical Waste Management in Algeria: A Review and Recommendations. Journal of Environmental Health Science & Engineering, 17(4), 2331-2342.
Health Care Without Harm (n.d.). Sustainable Healthcare.
Cleveland Clinic. (2021). Waste Management and Sustainability Initiatives.
Mayo Clinic. (2020). Waste Reduction and Recycling Programs.
Charité Hospital, Berlin (2022). Recycling and Waste Management Strategies.
Ministry of Health, France. (2021). Subsidies for medical waste management.

CONTENTS

Hospital Waste: Towards Sustainable Management in University Hospitals

By: G.L/ABU YOUCEF/2024

MIX
Papier aus verantwortungsvollen Quellen
Paper from responsible sources
FSC® C105338

Printed by Books on Demand GmbH, Norderstedt / Germany